HELPING INSURE OUR ENERGY FUTURE

A PROGRAM FOR DEVELOPING SYNTHETIC FUEL PLANTS NOW

A STATEMENT ON NATIONAL POLICY
BY THE RESEARCH AND POLICY COMMITTEE
OF THE COMMITTEE
FOR ECONOMIC DEVELOPMENT

JULY 1979

Library of Congress Cataloging in Publication Data

Committee for Economic Development. Research and Policy Committee.
  Helping insure our energy future.

  Includes bibliographical references.
  1. Synthetic fuels—United States.   2. Energy
policy—United States.
I. Title.
TP360.C65 1979   338.4'7'662660973   79-21027
ISBN 0-87186-769-9 lib'. bdg.
ISBN 0-87186-069-4 pbk.

First printing: July 1979
Paperbound: $4.50
Library binding: $6.00
Printed in the United States of America by Kearny Press, Inc.
Design: Stead, Young & Rowe, Inc.

COMMITTEE FOR ECONOMIC DEVELOPMENT
477 Madison Avenue, New York, N.Y. 10022
1700 K Street, N.W., Washington, D.C. 20006

CONTENTS

HELPING INSURE OUR ENERGY FUTURE

A PROGRAM FOR DEVELOPING SYNTHETIC FUEL PLANTS NOW

RESPONSIBILITY FOR CED STATEMENTS
ON NATIONAL POLICY

The Committee for Economic Development is an independent research and educational organization of two hundred business executives and educators. CED is nonprofit, nonpartisan, and nonpolitical. Its purpose is to propose policies that will help to bring about steady economic growth at high employment and reasonably stable prices, increase productivity and living standards, provide greater and more equal opportunity for every citizen, and improve the quality of life for all. A more complete description of CED is to be found on page 44.

All CED policy recommendations must have the approval of the Research and Policy Committee, trustees whose names are listed on page vii. This Committee is directed under the bylaws to "initiate studies into the principles of business policy and of public policy which will foster the full contribution by industry and commerce to the attainment and maintenance" of the objectives stated above. The bylaws emphasize that "all research is to be thoroughly objective in character, and the approach in each instance is to be from the standpoint of the general welfare and not from that of any special political or economic group." The Committee is aided by a Research Advisory Board of leading social scientists and by a small permanent professional staff.

The Research and Policy Committee is not attempting to pass judgment on any pending specific legislative proposals; its purpose is to urge careful consideration of the objectives set forth in this statement and of the best means of accomplishing those objectives.

Each statement is preceded by extensive discussions, meetings, and exchanges of memoranda. The research is undertaken by a subcommittee, assisted by advisors chosen for their competence in the field under study. The members and advisors of the subcommittee that prepared this statement are listed on page viii.

The full Research and Policy Committee participates in the drafting of findings and recommendations. Likewise, the trustees on the drafting subcommittee vote to approve or disapprove a policy statement, and they share with the Research and Policy Committee the privilege of submitting individual comments for publication, as noted on pages vii and viii and on the appropriate page of the text of the statement.

Except for the members of the Research and Policy Committee and the responsible subcommittee, the recommendations presented herein are not necessarily endorsed by other trustees or by the advisors, contributors, staff members, or others associated with CED.

RESEARCH AND POLICY COMMITTEE

[1]Voted to approve the policy statement but submitted memoranda of comment, reservation, or dissent or wished to be associated with memoranda of others. See pages 34-38.

NOTE/A complete list of CED trustees and honorary trustees appears at the back of the book. Company or institutional associations are included for identification only; the organizations do not share in the responsibility borne by the individuals.

PURPOSE OF THIS STATEMENT

Over a year and a half ago, the trustees of the Committee for Economic Development began to develop proposals for financing and construction of a first increment of commercial synthetic fuel plants. Concerned with the urgent need to increase domestic energy supplies and to reduce a growing dependence on imported oil, we brought together a highly qualified subcommittee of top executives and experts from energy companies, financial institutions, government, and construction.

When the CED Research and Policy Committee began exploring the prospects for the commercial development of synthetic fuels in 1978, the energy situation was serious, but relatively stable. But even at that time, we were gravely concerned about the uncertainties in cost and supply that continued to plague oil-consuming nations.

The cut-off of oil from Iran, the Three Mile Island accident, the gasoline lines of 1979, and the resulting public demand for action have only reinforced our belief that there is a critical need to increase domestic supplies. At last a consensus is emerging that the United States must free itself from excessive dependence on often insufficient sources of imported oil.

A CONTINUING INTEREST

This study has grown out of CED's long and continuing interest in energy. Before the oil embargo of 1973, CED began work on the policy statement *Achieving Energy Independence* (1974) which proposed measures to stimulate domestic production and reduce the growth of energy demand to bring about a high degree of U.S. energy independence. As one of our principal recommendations, we called for an intensified government effort to promote synthetic fuels.

The next CED statement on energy *International Economic Consequences of High-Priced Energy* (1975) presented an assessment of world energy and economic problems and recommendations to meet them by CED and six international counterparts. *Nuclear Energy and National Security* (1976) urged the United States to reassert its leadership in international nuclear affairs and to continue to develop its nuclear power industry. *Key Ele-*

x

ments of a National Energy Strategy (1977) called for an energy program both to increase domestic supply and to achieve greater conservation, placing primary reliance on the market system. Earlier this year, CED released an important new supplementary paper entitled *Thinking Through the Energy Problem* by Thomas C. Schelling. The paper concluded that energy costs could easily rise twice as fast as the average of other costs between now and the turn of the century. These higher costs should be reflected in higher prices since higher prices will make the greatest contribution to both conservation and increased supply.

THE ROLE FOR SYNTHETIC FUELS

Stepped up conservation efforts, increased oil and gas exploration, and continued development of nuclear power with proper safeguards should all be important elements in this country's energy future. Conservation, we believe, will prove vitally important, especially during the time it takes to make the transition from fossil fuels to the technologies of the next century.

But because both coal and oil shale are so abundant in the United States, we believe they will play an important role in overall U.S. energy strategy and that their development can go far in insuring a new national energy security. In addition, developing synthetic fuels can also bolster this country's international economic and political position by demonstrating to the world a new U.S. resolve to take charge of its energy future.

We are pleased with the recent interest that the Administration and Congress have expressed in developing synthetic fuels. However, we are concerned that in the rush to "do something" about energy, actions may be taken that would commit the United States to a huge, permanently-subsidized, government-run fuels industry. We believe that any such massive government program would be costly, inefficient, and worst of all—unnecessary.

RELYING ON THE PRIVATE SECTOR

The country can have the synthetic fuels it needs, and it can have them with a minimum of direct government involvement and cost. The results of our study indicate that if the government is willing to provide a limited number of financing incentives and to reduce certain environmental and regulatory barriers, the private sector is prepared to build and operate a number of plants that would demonstrate the commercial feasibility of converting coal and oil shale into synthetic oil and gas.

Placing primary reliance on the market system would be the most efficient and effective means of bringing these plants into commercial scale operation in the shortest time at the lowest public cost. While government action is needed, we also recommend that any government support program have firm dollar and size limits. Government should assume only those risks that are too large for private industry to assume alone. We are convinced that energy companies and financing institutions are ready to assume their full share of risks in building these plants as long as government is prepared to reduce the extraordinary regulatory and environmental risks as well as those associated with foreign supply not under industry control.

AN IMPORTANT CONSENSUS

Perhaps the single most important achievement this report represents is the consensus we were able to achieve among the extraordinary group of individuals who made up the subcommittee. The members, listed on page viii, represent wide experience and expertise at the highest levels of energy production, banking and finance, construction and government, including Richard L. Dunham and Joseph Swidler, former chairmen of the Federal Power Commission, and John C. Sawhill, former Federal Energy Administrator. We are also deeply grateful to a special group of key financial advisors led by Richard W. Manderbach, Senior Vice President, Bank of America, N.T.&.S.A.

We are especially indebted to subcommittee Chairman Roderick M. Hills, partner in Latham, Watkins and Hills and former Chairman of the Securities and Exchange Commission, for his leadership and for his sensitivity to the complex technological, economic, and political issues involved. We also wish to thank Frank W. Schiff, CED Vice President and Chief Economist and project director for the report, for his clear and insightful approach to this problem.

Franklin A. Lindsay, *Chairman*
Research and Policy Committee

CHAPTER 1
SUMMARY

This Committee believes that there is an urgent need for the United States to increase its domestic energy options, thereby lessening the nation's dependence on often insecure foreign energy supplies.

Recent events in the Middle East and the apparent setback to nuclear power development caused by the Three Mile Island incident provide dramatic evidence of the strategic and economic vulnerability of the United States to sudden curtailments in energy availability.

In the near-term future, much more must be done to *conserve energy* and to *expand the output of existing types of domestic energy resources.* We reaffirm our long-standing conviction that the most important step toward achieving these goals is early action to foster maximum feasible reliance on free market forces—by letting the price system work effectively, by expediting environmental and other regulatory decisions, and by removing unreasonable regulatory requirements.

This kind of action will not only encourage expanded output from existing energy sources but will also spur development of new sources. However, our national needs make it imperative that we also do more now to *accelerate the introduction of new energy technologies* that can serve the nation in the medium and longer-term future.

We welcome the emphasis which the President's recent energy messages place on encouraging new initiatives to foster enlarged domestic energy supplies. Our aim in the report is to present specific proposals for carrying out such initiatives in a way that will make optimum use of the strengths of the private market system and of private financial resources.

FOSTERING SYNTHETIC FUEL PRODUCTION

This statement focuses on one element of the multi-faceted strategy needed to deal with the energy problem over the medium-term future (seven to twenty years ahead): the development of an initial capability for commercial-scale domestic production of synthetic gas and liquid fuels from coal and shale.*

Coal gasification and the extraction of liquid fuel from both coal and oil shale constitute two of the most promising alternative energy sources for the medium-term future. Adequate technological knowledge of these processes is already available or (in the case of coal liquefaction) is likely to become available soon. What is needed are demonstrations of commercial feasibility. Both coal and oil shale are present in massive quantities in this country, and synthetic fuels from these sources can be used to supplement domestic production of petroleum, natural gas, coal, and nuclear power.

Because they can provide fuel for vehicles and other types of transport, synthetic fuels are of particular value to an America dependent upon a vast transportation network. Furthermore, the commercial-scale production of synthetic gas from coal can help replace petroleum-derived products in many applications, thereby making it possible to reserve more domestic petroleum for transportation.

The immediate need, in our view, is to develop a limited number of commercial-size synthetic fuel plants (in the neighborhood of ten such plants) that would be sufficient to demonstrate the technology, environmental effects, and economic viability of various synthetic fuel options.** This will provide the base for a subsequent rapid expansion in the number of such plants if they are needed.

This effort should place primary reliance on the technical and financial resources of the private sector and on the incentives provided by market forces. However, because of the unusual risks involved, the distortions created by past regulatory actions, and the uncertainties posed by potential future governmental actions in the environmental and other areas, some added governmental action will also be needed. Without such governmental assistance, the prospective return on investment may not produce sufficient incentives to potential investors to bring about the needed early development.

*See memorandum by CHARLES P. BOWEN, JR., page 34.

**See memorandum by GEORGE C. McGHEE, page 35.

The proposed construction of these plants might entail total costs to the private and public sectors in the order of $15 billion without allowance for future price inflation.[1]*However, government would pay only part of these costs, perhaps no more than around $2 billion spread over the next decade or more.** Of course, future events are unpredictable enough so that the final costs could be markedly higher or lower than this tentative figure.

We believe that the cooperative public-private approach outlined in this statement is justified because of the benefits that early development of the needed plants will confer upon the nation as a whole.

In essence, this kind of government action now can be viewed as a form of national insurance against the severe potential risks of future energy curtailments. Some argue that the cost of government assistance in this area is both unnecessary and unwarranted, especially in a time of tight budgets and taxpayer revolts. But while such costs are likely to be substantial, we believe that the costs of not proceeding with the development of new energy technologies could be even higher. The longer action is delayed to initiate the kind of market-oriented synthetic fuels program we recommend, the greater the likelihood that the country will be propelled into a huge, fully government-funded crash program to do the same thing that our proposal can achieve more efficiently and at a lower overall cost.

EXISTING POLICY OPTIONS

There are a number of techniques government could use to assist the development of commercial-scale synthetic fuel plants. We need, first, to take note of some of the options that already exist or that have been under active consideration for some time.

One such incentive is the new *additional 10 percent investment tax credit* that applies to such energy projects. For this measure to be effective, however, the period in which this credit can be taken needs to be extended significantly beyond its present 1982 cut-off date.

A second kind of incentive was recently proposed by the President: provision for a federally-financed $3 per barrel *oil shale production tax credit.*

A third incentive could be applied without new legislation but requires approval by the Federal Energy Regulatory Commission (FERC) before it could be implemented in practice: the so-called *all-events tariff.*** This would only be applicable to companies subject to utility-type regulation and would not entail any direct costs for the federal budget, although it

---

[1]/ These estimates exclude the cost of associated coal mines.

---

*See memorandum by JACK F. BENNETT, page 35.

**See memoranda by G. BARRON MALLORY, pages 35 and 36.

would, of course, entail costs to the ultimate consumers. It would provide assurance by FERC that the burden of debt incurred in the construction and operation of proposed synthetic fuel plants (plus an allowed return on equity invested) can be automatically passed on to consumers in the form of higher rates.

ENLARGING THE FINANCING AND INCENTIVE OPTIONS: PRINCIPLES FOR PUBLIC-PRIVATE COOPERATION*

We believe that added financing and incentive options are needed. No single solution will fit all circumstances. In devising our proposals for new options, however, we have been chiefly guided by the following general principles:

- Construction, ownership, and operation of new energy facilities should be in private hands, as much as possible, to encourage greater economic flexibility and to provide stronger incentives for efficient operation.

- Any government effort to aid in the development of new energy resources should be designed in a way that:

—causes the least interference with competitive pricing mechanisms and efficient resource allocation,

—permits identification of the cost of incentives provided by the government, and

—achieves the desired results in the most efficient fashion and at the lowest cost.

- Government should assume only the risks and costs that cannot be assumed on strictly commercial grounds, such as those caused by the government's price restraints, regulatory delays, and changes in environmental rules, and by actions of foreign energy-expanding countries or cartels.

- There should be adequate safeguards against poor management and positive incentives for efficient performance. Governmental intervention in internal operations should be avoided to the extent possible. At the same time, government should intensify its efforts to eliminate existing interferences with free market forces.

- Government support should not be biased to give undue competitive advantage to one type or size of firm.

- There should be firm dollar limits on the obligations assumed by the U.S. government under any program of government assistance for new technology energy plants.

*See memorandum by ROBERT R. NATHAN, page 36.

ENLARGING THE FINANCING AND INCENTIVE OPTIONS: RECOMMENDATIONS

We propose that the President be empowered to offer certain new financing options and incentives for the development of a limited number of first-of-a-kind commercial-scale synthetic fuel plants:

- long-term guaranteed price and purchase contracts
- completion undertakings
- and, within limits, debt guarantees.

Completion undertakings would provide specified protection against the risk that the project might not be completed because of cost overruns caused by changes in specifications, regulatory delays, or other governmental intervention. Such undertakings would, where appropriate, be used in conjunction with guaranteed price and purchase contracts. Debt guarantees represent an alternative option that would only be employed in certain limited circumstances. All incentives should be administered in a way that holds the total governmental assistance for individual projects to the lowest cost consistent with accomplishing the basic objective.

Such added financing and incentive options, together with the proposed changes in environmental procedures described below, would minimize the special risks and costs now associated with these projects. In our view, it is these special risks—rather than lack of an adequate supply of private funds—that constitute the principal deterrent to investment in first-of-a-kind synthetic fuel plants.

LONG-TERM GUARANTEED PRICE AND PURCHASE CONTRACTS. Under such a contract, the government would agree either to purchase a set volume of the product of the synthetic fuel plant at a guaranteed price for a specified period of time or to pay all or part of the difference between the agreed price and the regulated or market price for a fixed number of years, up to an agreed maximum amount (for example, sufficient to cover amortization of long-term debt). In contrast to the investment tax credit or the alternative production tax credit, which provide a fixed subsidy no matter what the market price is, the guaranteed price contract only obligates the government to pay the spread if the market price is, in fact, below the contract price.

This approach should incorporate some form of competitive bidding by companies or groups of companies that would vie for the right to negotiate a final contract. The lowest bids would presumably be above current market prices for fuel but all the incentives created by the market would, nonetheless, be present. Thus, the guaranteed price contract provides management with strong incentives to control construction and operating costs. Moreover, when production starts five to ten years hence, the market price might equal or exceed the guaranteed price, thereby eliminating the government subsidies. This was the result of the guaranteed price and purchase contracts used to speed the production of aluminum in World War II.

COMPLETION UNDERTAKINGS. In some cases, adequate financing may not be possible unless lenders are also given a guarantee against the risk that the project might not be completed because of such factors as excessive construction cost overruns caused by regulatory delays and other government interventions. To furnish some measure of protection against this contingency, the government should, in carefully defined circumstances, provide limited completion guarantees (for which a fee would be charged to the company) covering a specific number of new synthetic fuel plants with a specified maximum capacity.

DEBT GUARANTEES. There may be some circumstances in which well qualified firms cannot take sufficient advantage of the other options cited and lack a strong enough equity base to enter bids (for example, certain regulated companies). In such cases an outright government debt guarantee for at least certain parts of the project may be needed to induce adequate financing.

It should be noted that completion undertakings and debt guarantees, if used at all, would not normally protect against the loss of the equity committed to the project.

In conjunction with long-term purchase contracts, completion undertakings, and debt guarantees, the government should retain the option to buy out and close down the facility, under defined conditions, as an alternative to paying for its completion.

IMPROVING ENVIRONMENTAL PROCEDURES

While environmental concerns must be taken into account in connection with all energy technologies, we believe it is important to find workable solutions for the environmental problems that may be encountered in the construction and operation of new synthetic fuel facilities. Our proposals are designed to achieve a balanced trade-off that will allow a limited number of first-of-a-kind synthetic fuel plants to be built quickly, without any sweeping, across-the-board changes in environmental procedures.

For environmental relief, we recommend that the following powers be granted to the President:

1. Authority to designate a limited number of specific potential sites for synthetic fuel plants.

2. Authority to expedite governmental decisions concerning synthetic fuel plant operations in these areas.

3. Limited authority to grant variances from certain environmental standards and related pollution control requirements for these sites when national interests are overriding and when all reasonable alternatives to full compliance with existing environmental regulations are judged unworkable or excessively time-consuming. We are especially concerned about those cases in which a plant may be designed to meet all existing environmental standards, but where there is still uncertainty as to whether the plant will perform according to that design, the standards will stay unchanged, or other pollution sources external to the plant might combine to produce a noncomplying situation. Such uncertainty could make the financial risk of proceeding far too high even though the plant may, in fact, meet all standards when it is completed.

To safeguard regional and state interests, we also propose that presidential decisions with respect to synthetic fuel plant development or operation at such specific sites be subject to override by legislative act of Congress or by the state concerned, within a limited time period. We also believe it is important that the federal legislation for synthetic fuel development be accompanied by companion legislation in the various states where plant sites are expected to be located.

As part of the proposed authority, the President should be allowed to place limits on the time taken for action on environmental impact statements, on the time taken to issue needed permits, and on the time in which litigation opposing such permits can be brought. In addition, he should have authority to expedite the appellate process.

We also recommend that, once a plant is in operation, it should not be required at a later time to install more stringent control technology at its own cost. If it is determined that more stringent controls are essential in the public interest, the cost of such retrofitting should, for this very limited number of designated plants, be borne by the government requiring it.

* * *

If the governmental actions proposed in this statement are instituted, there is strong reason to believe that it will be possible for the private sector to construct and operate the limited number of commercial-scale synthetic fuel plants so urgently required by the national interest.* We are convinced that both energy companies and financing institutions are ready to play their full part in risk sharing and credit extension as long as the government is prepared to reduce the extraordinary risks not under industry control.

---

*See memorandum by C. WREDE PETERSMEYER, page 37.

CHAPTER 2
INTRODUCTION AND BACKGROUND

The Committee for Economic Development has long been concerned with the energy future of the United States.

Our most recent policy statement in this area, *Key Elements of a National Energy Strategy* (1977), asked that U.S. energy policies be guided by certain fundamental principles: a balanced program of both increased energy supply and conservation; maximum reliance on the market system; greater use of coal and nuclear power; and accelerated research on new energy technologies.

We reaffirm our support of these principles. (See page 9.)

In our recent statements on this subject,[1] we stressed the urgent need to follow these principles in order to reduce U.S. dependence on foreign energy sources and to minimize the impact of increases in real energy costs. Since then, the need to deal effectively with these issues has become even more pressing, but public understanding of the problem and U.S. government actions still fall far short of what is required.

There have been reports of significant discoveries of new energy supplies, but most of these discoveries have been outside U.S. borders. Events in Iran and elsewhere have underlined the extent to which U.S. dependence on foreign energy supplies can subject the economy to serious risks of disruption. In our view, these events sharpen the overriding need for prompt and effective public-private action to provide a higher degree of national

---

[1] Preceding CED policy statements in the energy field were *Achieving Energy Independence* (1974), *International Economic Consequences of High-Priced Energy* (1975), and *Nuclear Energy and National Security* (1976). In April 1979 a CED supplementary paper, *Thinking Through the Energy Problem,* was prepared by Thomas C. Schelling, working with the CED Design Committee on Long-Range Energy Policy.

In CED's 1977 policy statement, *Key Elements of a National Energy Strategy*, we urged that national attention be focused on the basic principles involved in developing a sound energy policy.

Now, two years later, we are concerned with the slow pace of energy policy development. We feel there is a very real danger that the discussion over our energy policy is becoming so enmeshed in details and specifics that fundamental considerations are obscured.

Therefore, we reaffirm those basic principles:

● PROMOTING INCREASED SUPPLIES AND GREATER CONSERVATION. Equally strong emphasis must be placed on promoting increased supplies of currently usable fuels and on encouraging greater conservation.

● RELYING ON THE MARKET SYSTEM. Public policy should rely primarily on market incentives in the production, distribution, and consumption of energy supplies. Government actions should be a supplement to, not a substitute for (or obstacle to) market forces. Transitional measures needed to avoid major disruptions in the economy must be kept temporary.

● INCREASING RELIANCE ON COAL AND NUCLEAR ENERGY. The nation's interim aim should be to adjust to eventually smaller supplies of oil and gas by increasing its coal and nuclear power capacity. The nation may have to rely on coal and nuclear power for many years before these energy sources can be supplemented or replaced economically by advanced energy technologies.

● DEVELOPING ENERGY TECHNOLOGIES FOR THE NEXT CENTURY. Work must be accelerated now on advanced nuclear technologies, solar energy, and other new energy technologies so that the nation will be ready for the time when fossil fuels become seriously depleted.

● RESOLVING CONFLICTS BETWEEN ENERGY AND THE ENVIRONMENT. Better and more expeditious ways must be found to settle conflicts that arise between the requirements for increased energy supplies and for environmental protection. The aim should be to avoid serious and long-lasting damage to elements of the environment while permitting energy resources to be exploited expeditiously.

● STRENGTHENING INTERNATIONAL COOPERATION. Energy is an international problem. U.S. actions must give full consideration to the interdependence of consuming and producing countries and must take into account America's vital interest in the energy and financial security of its friends and allies.

insurance against the risk of major future shortages of reasonably-priced energy supplies.

In the next five to seven years, U.S. dependence on foreign energy sources can only be alleviated by intensified efforts in two areas: (1) energy conservation, which has the potential to provide the fastest relief; and (2) further expansion of existing types of U.S. domestic energy resources. Starting in the mid-1980s, however, a progressively wider range of options can gradually become available *if* proper steps are taken now to:

- accelerate oil and gas exploration
- devise more effective means of energy conservation
- facilitate further use of nuclear power under suitable safeguards
- develop geothermal energy
- experiment further with solar energy and new technologies for obtaining heat and power by direct burning of our massive coal reserves, and
- bring such synthetic fuel technologies as coal gasification and liquefaction, production of alcohol fuels from coal and agricultural and forest products, and extraction of oil from shale to the stage of commercial production.*

In light of the great uncertainties surrounding projections for the discovery of oil and gas deposits in this medium-term (seven to twenty year) period and of the additional uncertainties regarding the pace of medium-term nuclear development that have arisen as a result of the Three Mile Island incident, we believe there is a pressing need to develop and keep open an adequate number of the other energy options.

The long lead times involved in developing such new sources and technologies, and in actually constructing new types of energy facilities, reinforce the need for prompt action to enlarge the range of available energy supply options for the medium-term period, particularly where these options can utilize the capacity of the private sector and relieve the pressures for governmental programs. Indeed, the longer such potential private programs are delayed, the more impelling will it become to turn to huge, fully government-funded crash programs to deal with an energy crisis that has not been adequately anticipated and prepared for.

This policy statement sets forth specific proposals that would better equip the federal government to start moving the United States *now* toward greater self-reliance in one promising source of medium-term energy: synthetic fuels.

The statement does *not* support the creation by the government of a permanently-subsidized large-scale synthetic fuels industry. Rather, it calls

---

*See memorandum by W. D. DANCE, page 37.

for early action — with government support but with primary reliance on private sector participation — to build a small number of first-of-a-kind operating plants that could demonstrate the commercial feasibility of converting coal and oil shale to synthetic fuels.

Our proposals call for selective removal of various regulatory and other governmental disincentives to such action, based on a careful balancing of energy-related and other national objectives. This includes, in particular, the elimination of undue regulatory delays and uncertainties. In addition, the proposals call for some positive governmental incentives. The central emphasis of these recommendations is on using the technical, managerial, and financial resources of the private sector with maximum economic efficiency, placing the greatest feasible reliance on market forces, and minimizing net additional expenditures of government funds.

We believe that the proposed program can and should be carried out in a manner that is consistent with overall fiscal and budgetary discipline. However, if the cost of the measures we recommend should turn out to be more than can be accommodated by offsetting savings elsewhere in the budget, then we would support action to cover the costs of these programs by providing for a correspondingly higher level of taxes than would otherwise be the case.

THE RATIONALE FOR SPECIAL MEASURES: INSURANCE FOR THE FUTURE

Our present energy problems, serious enough in an inflationary world economy faced with limited petroleum resources located largely in politically volatile areas, have been greatly aggravated by three factors:

(1) regulatory distortions of energy pricing which have led to excessive consumption and lagging production

(2) uncertainties about the cost and availability of energy from alternate sources, including synthetic fuels, and

(3) uncertainties about future governmental regulations, especially those relating to price and the environment.

We are convinced that the foremost principle for dealing with energy over the coming decades is to let the price system work. If energy prices in the past had been allowed to rise to levels determined by market forces, there would be far less need now for even the limited government assistance we propose. The demand for a stronger national energy policy thus must not become a demand for federal intervention in all energy decisions.

But given the crippled pricing system now at work in energy along with various other current impediments to synthetic fuel production, it is unlikely that a sufficient number of the needed commercial-scale synthetic fuel plants will be built by the private sector in time to help protect the U.S. economy from the serious continuing threat of worldwide energy shortages.

There are a number of basic reasons why this is so.

First, governmental policies that have distorted the economics of energy production and conservation in the past are now discouraging needed investment in new technologies. Significant time will be needed to counterbalance these distortions.

Second, the risks and future uncertainties of launching synthetic fuel projects now are greater than most private investors can accept without some form of government assistance or special action.

Foremost among these risks are the uncertainties about future regulatory practices. Such practices can alter the plant cost or selling price of an energy product enormously and, particularly for environmental reasons, can bar a plant's production indefinitely.

Also, some current estimates suggest that the cost of oil or gas produced by such plants, if they were to become operational now, might be one and one-half to two times as high as the current costs of production from conventional sources. While the rising costs of existing energy sources could substantially reduce (or even eliminate) this anticipated cost differential by the time the plants become operational, uncertainties about relative future costs remain an important impediment to investment in the new-technology plants.

In any case, the cost of each individual plant promises to be very high, perhaps ranging up to $1 or $2 billion in 1979 dollars. Finally, the technologies involved are fairly complex and even though they have been tested in pilot and demonstration plants, they have not been subjected to the conclusive tests of commercial-scale installations in the United States.

In the face of these risks and uncertainties, private U.S. investors will have to go slow, if they go at all, on investing in new high-cost technologies. Thus, special governmental incentives may be required to induce firms to undertake the needed investments now or in the near-term future. Such incentives should be used in conjunction with—not as substitutes for—future removals of price controls and governmental actions to achieve more rational and timely regulatory decisions, particularly in the environmental area.

Since it is the declared policy of the federal government to bring synthetic fuel production to the commercial development stage as early as possible, it is appropriate that the government reduce some of the costs and financial risks involved in the private development of these new energy technologies. **Such action is desirable when it can be expected to secure greater benefits to the nation as a whole than the benefits that would accrue to private investors through the workings of the market system alone.** By the same token, the costs of *not* undertaking the needed investments can be much greater for the nation than for individual investors.

- Such investments could be a major factor in reducing the dependence of the United States and its friends and allies on insecure overseas energy supplies.[2] They would demonstrate our determination to curtail this dependence. Even though world energy supplies may turn out to be adequate for the next ten or twenty years, there is a real need to develop the capability to make a major shift from foreign to domestic energy supplies in order to adequately protect the national interest.

  Today the United States is importing close to 50 percent of the oil it consumes, and the demand for foreign oil and natural gas is likely to increase. This exposure to interruption of U.S. overseas energy supplies is a significant military and economic risk. We believe our country simply must not allow these trends to continue.

- Even if strict market conditions did not warrant the project at the present time, beginning work now on these projects is essential for developing a broader base of knowledge and for reducing uncertainties about the nature of the medium-term energy problem. The information to be gained from these investments will help both government and industry estimate and plan for the energy problems that will face this country into the 1990s and the next century.

- A proven U.S. capability to produce synthetic fuel on a commercial scale, and the knowledge that such production could be expanded, should inhibit possible longer-run attempts by foreign producers to raise their prices above the estimated cost of such synthetic fuels.

- Furthermore, an actual increase in domestic energy production and conservation, which helps reduce imports, is likely to have a dampening effect on the world price of oil and reduce the average price of U.S. oil imports. Conversely, the true "cost" of imports exceeds the price paid per barrel of oil.[3]

---

[2] The interests in energy matters of the United States and its key friends and allies abroad are closely linked. This fact is recognized in formal commitments of the United States to share oil with such friendly foreign nations in the event of major disruptions of oil imports.

---

[3] For a more detailed discussion of this point, see Thomas C. Schelling, *Thinking Through the Energy Problem* (New York: Committee for Economic Development, 1979), pp. 29-30.

A central theme underlying all these arguments for some form of government assistance to the private development of synthetic fuel production is the need for adequate national insurance against further severe strategic, economic and balance-of-payments damages that the United States could suffer as a result of major future curtailments of energy availability from foreign sources. Such curtailments would release another inflationary force in the world and could result in shortages of energy that could impede economic activity, further increase inflation and unemployment, reduce real standards of living, and discourage investment in future productive capacity.

FOSTERING ALTERNATIVE ENERGY SOURCES

For all the foregoing reasons, there is a strong case for a carefully considered and closely limited program of governmental encouragement of alternate energy sources.

Research and development efforts on potential renewable sources with least environmental impact, such as solar and fusion energy, are being pursued with substantial government assistance. But large-scale supplies from such sources are probably still decades away. Therefore, there is also a need to implement the commercial production of medium-term domestic synthetic energy sources which can begin to supplement or replace oil and gas over the next ten to fifteen years.

Four abundant energy sources that can be significantly expanded over the medium term are coal, oil shale, tar sands, and nuclear energy. Currently, the most promising medium-term fossil-based energy sources are coal gasification and the extraction of liquid fuels from coal and oil shale. Some of the technologies for doing this are ready for a final verification in commercially-sized installations. Others may be ready in the near future. Both coal and oil shale are abundant in this country and synthetic fuels from these sources can be used to supplement domestic production of petroleum and natural gas. In particular, these synthetic fuels in liquid form can be used in place of petroleum products to power our military, commercial, and personal transportation.[4]

Accordingly, this statement concentrates on what government and the private sector can do to demonstrate this country's capacity to produce synthetic fuel from coal and oil shale on a commercial basis. While our proposals focus on synthetic fuels, some or all of the proposals we present may also deserve consideration in connection with various other new energy technologies when these reach a similar stage of development.

4/ In turn, improvements in transportation facilities could make a major contribution toward increased production and use of coal.

KEY INGREDIENTS IN AN ACTION PROGRAM

Because projections of future world energy supply and demand can have a substantial margin of error, no one is sure precisely how many synthetic fuel plants may be needed. What seems certain, however, is that the national interest would be well served if the Executive Branch were given the authority now to facilitate the building of a limited number of such plants of minimum commercial size to demonstrate concretely the costs, environmental impact, and productive capacity they involve. Such plants would represent a base which could subsequently be expanded rapidly by the private sector if cost and price developments make it commercially attractive. Furthermore, there will be a pool of trained engineers and managers and a body of operating experience which will allow rapid acceleration of the construction of other plants if the need arises.

What is needed is joint action by business and government. Business needs to indicate that it is ready and willing to work with government to provide capital and management to start up high-risk, high-capital energy projects. By working together and by sharing the enormous risks involved, business and government can help to assure the United States a more viable energy future.

The cost of these new energy projects will inevitably be large. Because most of these projects cannot currently be constructed with an assurance that the initial output will be saleable at world market prices, although this is not necessarily ruled out, a significant initial element of government support may very well be needed not only to cover a portion of the construction costs but also to cover a long-enough period of operation to pay off the debts incurred in building the project.

Thereafter, whether any of these first-of-a-kind plants would continue to operate would depend upon whether the prevailing technological cost, price and market situation made such operation worthwhile. The program we envisage does not countenance the building up of a series of plants to be kept running indefinitely by means of permanent operating subsidies.

Any such program must be fair and equitable to both the public and the private interests. It must not be a program under which government assumes all of the risks while industry reaps the profits. As far as possible, all participants must share in both the downside risks as well as the upside rewards.

CHAPTER 3
A PROPOSED FINANCING APPROACH

For the reasons stated in Chapter 2, we believe there is now an urgent need to empower the federal government to provide new assistance to the private development and construction of a limited number of first-of-a-kind, new-technology domestic energy facilities.[1]

EXISTING POLICY OPTIONS

Before turning to our own proposals for additional synthetic fuel financing options, we shall first take note of an incentive that now exists (the energy investment tax credit), one that was proposed in the President's energy message of April 5, 1979 (the shale oil production tax credit), and one that is under consideration by the Federal Energy Regulatory Commission (FERC) and that could be implemented without new legislation:

(1) The Energy Tax Act of 1978 provides a substantial investment incentive in the form of an additional 10 percent investment tax credit for designated new energy projects (including synthetic fuel projects).

---

[1]/ This policy statement is directed at efforts to bring synthetic fuel production to a commercial stage after the feasibility of the basic technology has been demonstrated. CED's recommendations for increased government assistance to basic research and development efforts are contained in a forthcoming policy statement, *Revitalizing Technological Progress in the United States*.

The *energy investment tax credit* is primarily of value only to companies profitable enough to enable them to use the tax credits that can be generated by the very large investments required for new energy projects. This credit will enlarge the cash flow of such companies when they make capital expenditures, thus strengthening their financial capacity. Long-term lenders may be willing to extend credit for new energy projects to such companies without requiring government guarantees if these firms will put their financial credit behind such projects. For many other companies, however, the energy tax credit may provide little incentive.

Also, the energy investment tax credit expires in 1982. **Yet the kind of new synthetic fuel plants that are needed will, in most if not all cases, take years longer to construct. Accordingly, if the tax credit is to be effective at all, the period in which it can be taken should be extended.**

(2) To permit the demonstration of commercial shale oil production by the mid-1980s, President Carter has recommended that part of the proceeds of his proposed Energy Security Fund be used to finance the costs of a limited $3 per barrel shale oil production credit. This credit would begin to phase out when world oil prices reach specified levels. It represents a way of offsetting at least part of the currently expected excess of shale oil production costs over those of competing oil sources.

(3) The proposed *all-events tariff** is applicable only to the regulated part of the energy industry. It would provide no direct federal assistance, but would require federal and/or state action to assure that the full burden of servicing the debt arising from the construction and subsequent operation of synthetic energy facilities (plus an allowed return on equity) will be automatically passed through to specified gas transmission and distribution companies and, ultimately, to final consumers in the form of higher rates.[2]

This "tariff" would apply only where the fuel is subject to rate regulation, so it is primarily relevant to the construction of plants that will produce high BTU gas for pipeline transmission to utilities for retail distribution. It could be established today by FERC, with similar action by state regulatory bodies.[3]

---

[2] Under previously existing legislation, FERC has authority to establish such an all-events tariff for interstate gas. The 1978 energy legislation extends this authority to intrastate gas as well. Technically, the authority extends only to *natural* gas. However, this authority can be extended to synthetic gas plants to the extent that the output of such plants is commingled with natural gas.

---

[3] While it does not appear that conforming action by state regulatory commissions would be necessary to make the FERC all-events tariff order legally effective (at least as it relates to the pass-through of costs from the gas producer to the regulated utility), lenders may, before committing their funds, want to be assured of a cooperative view of the matter by such commissions to remove any doubts as to whether the charges will be fully passed through to the final consumer.

---

*See memoranda by G. BARRON MALLORY, page 36
and by ROGER B. SMITH, page 37.

A specific proposal is now under formal consideration by FERC that calls for establishment of an all-events tariff to permit the financing of a coal gasification plant by a consortium of five large gas distribution companies.[4]*

The all-events tariff could be instituted relatively quickly, since the financing costs and (depending on the particular tariff used) all or part of the risk of noncompletion of the plant are borne by consumers. Moreover, the cost to individual users is likely to be relatively low.[5] Such costs could be spread over a relatively large number of gas consumers, although they would not be borne by all energy users.

The all-events tariff will only be effective if there is assurance that FERC and the state regulatory bodies will permit the pipelines and distributing utilities to earn an adequate rate of return on the operation and that the rules will not change.

Also, any order establishing the all-events tariff would have to provide adequate safeguards against poor management and incentives for superior performance.

DEVELOPING ADDITIONAL OPTIONS: UNDERLYING PRINCIPLES

We think it is apparent that other financing options are needed. Those that we here propose are consistent with the following underlying principles:

---

[4] The proposal was turned down by an administrative law judge in June 1979, but this ruling is subject to review and final action by the full Federal Energy Regulatory Commission.

Under the proposal, construction loan financing costs plus a return on equity during the construction period would be passed on, as incurred, in the form of rate increases to the customers of the five sponsoring distribution companies. Authority would also be provided through "rolled-in pricing" to spread the higher costs uniformly across all customers of the five companies. After completion of the plant, debt service, operating costs and an allowed return on equity would be automatically included in the gas rate structure. In the event the plant is not completed, funds advanced by the lenders and part of the equity money already invested would be repaid over a five-year period by means of a special surcharge to the customers of the five utilities.

Full or partial equity refunds would be limited to cases in which noncompletion is due to such factors as major unexpected changes in government regulations. It would not be available in instances where abandonment of the project is due to technological factors or inefficient management (though such conditions, in some cases, may not be easy to define or administer).

---

[5] This appears to be true under the already pending proposal involving five large distribution companies. The companies estimate that even if the project fails, payments to lenders up to a total of $1.2 billion would cost consumers of the five companies, who buy gas for home use, no more than $15 per annum, on the average, over five years. It may also be feasible to combine special government incentives of the kind described below with an all-events tariff. This would make it possible to pass on at least part of the benefits of the government subsidy involved to the ultimate consumer in the form of a smaller rate increase than would otherwise have to be instituted.

---

*See memorandum by J. W. McSWINEY, page 37.

• New forms of domestic energy capacity should be developed in a manner that is least likely to:

> (1) interfere with competitive pricing mechanisms and distort the efficient long-term allocation of resources among alternative forms of energy production and efforts to foster conservation, or
>
> (2) create subsidies that would give continuous support to inefficient technologies or inefficient producers.

• New programs should be designed to enhance our ability to know the "real" costs of alternate fuel sources so that we will have a basis for informed decisions on further expansion of competing technologies.

• Where there are alternate forms of government support for a specific type of energy capacity, they should be limited to the problems of raising capital and should not result in large subsidies for inefficient producers. In fact, the relative amounts of government support to individual firms should avoid giving undue competitive advantage to one type or size of firm.

• Ultimately, the costs of energy should be borne by the consumer to promote efficient energy use and make conservation effective. However, where accelerated development of a new energy source is deemed in the national interest, the additional costs should be borne by the nation at large through the tax system or through charges that are borne equally by all energy users.

• Any government effort to develop new energy resources should be designed to:

—permit identification of the cost of government incentives, and
—achieve the desired results in the most efficient fashion and at the lowest cost.

• Construction, ownership, and operation of new energy facilities should be in private hands to encourage greater economic flexibility and to provide stronger incentives for efficient operation.

• Government should assume only the risks and costs that cannot be assumed on strictly commercial grounds, such as those caused by the government's actions in terms of price restraints, regulatory delays, revisions in specifications, or changes in environmental rules. Adequate safeguards against poor management and incentives for efficient performance must be preserved and governmental intervention in internal operations should be avoided to the extent possible.

• There should be firm dollar limits on the obligations assumed by the U.S. government under any program of government assistance for new technology energy plants.

In devising appropriate combinations of incentives that conform to the above principles in different circumstances, a number of considerations must be taken into account.

For example, equity holders should be expected to bear the risks associated with plant construction and producing the final product once the technology is adequately developed, needed raw materials are available, and the price for which the product is to be sold is largely free of government regulation. But business firms cannot assume all of the unusual risks associated with uncertainty about the market prices that may arise from governmental interventions (whether because of environmental concerns or of those regulatory requirements that cause delay). Nor should these firms bear all of the added technological and economic risks and costs that can arise if the timetable for completion of the new facilities is accelerated because of national interest considerations.

Lenders can only assume normal credit risks and cannot be expected to assume risks related to actions by governmental entities that impinge heavily on major new technologies. Government should not be expected, however, to guarantee lenders against all the risks related to synthetic fuel projects. Firms undertaking new projects should be in a strong enough position in terms of their capitalization, technological and production capacity, and access to raw materials to protect their lenders against many of the risks.

Finally, new government programs should distinguish between energy projects that will deliver products subject to governmental rate setting (for example, by utility-type regulation) and those projects that will deliver products at market-determined prices.* In the former case, the risks facing the project are more related to regulatory decisions affecting the allowable rate of return, while in the latter, more risk attaches to the possibilities for fluctuations in market prices.

RECOMMENDATIONS FOR ALTERNATIVE POLICY OPTIONS

1. GUARANTEED PRICE AND PURCHASE CONTRACTS

We believe that the President should be empowered to offer guaranteed price contracts to encourage prospective builders of a limited number of synthetic fuel plants. **Specifically, we propose that the government invite bids and enter into long-term guaranteed price contracts with potential participants in a program for accelerated development of first-of-a-kind commercial-scale synthetic fuel plants. The government would agree either to purchase a set volume of the product of the synthetic fuel plant at the guaranteed price for a specified period of time, or to pay all or part of**

---

*See memorandum by JACK F. BENNETT, page 38.

HOW A GUARANTEED PRICE CONTRACT WORKS

Under the procedure we propose, the government would invite bids for the provision of a specified total quantity of a specified type of synthetic fuel. Separate bids would be solicited for different types of fuel and, where appropriate to test alternative production processes, for fuels produced by such different processes. Thus, one type of contract could be let for a given quality of pipeline gas produced from coal, while another contract might be for oil produced from shale through an *in-situ* process. An alternative procedure that might also be considered in some cases would be to solicit bids for the specified end product only, thus allowing for competition among different processes.

Firms participating in the bidding process would be eligible for the various alternative forms of government assistance we have described. Bidding firms would be asked to specify the extent of government assistance they intend to utilize (of the types provided under this synthetic fuel program). That information would then be taken into account by the administering agency in deciding which bid represented the lowest probable cost to the government.

If under a guaranteed price contract, the market price turned out to be higher than the minimum guaranteed price, the firms would be free to sell their product at such market prices as they wished. In this instance, the government would incur no costs and might, under the contract, actually share the incremental revenue with the producer. If the market price fell below the minimum guaranteed price, the government would make up the difference by purchasing the product for its own use at the guaranteed price or by letting the firm sell the product in the market and reimbursing it for the differential between the open-market and the guaranteed price.

The type of guaranteed price contract described here has numerous advantages. It is relatively simple and straightforward. It does not add to government interference with market prices, and the extent of a government subsidy, if any, is clearly apparent. Also, the government does not get involved in the internal operations of energy facilities or in the actual design of the facility. Management incentives for efficiency and holding down costs are preserved under the competitive bid procedures; thus, the firm still bears all (or under some contracts, a part) of the burden of cost overruns above projected levels. The arrangement can be structured so that investors and the government share in the "upside" as well as "downside" risks. From the point of view of lenders, the arrangement may be viewed as a more reliable government commitment than various other alternatives since it involves a formal long-term contract at a specified price.

the differential between the agreed price and the regulated[6] or market price for a fixed number of years, up to an agreed maximum amount. The guaranteed price would be established on the basis of competitive tenders for a specified total volume of a specified type of synthetic fuel (for example, pipeline-quality gas produced from coal). The contracts would be negotiated with the firms that proposed the lowest support price levels. Separate bids would be invited for different types of synthetic fuels.

The guaranteed price might include an inflation adjustment. Producers involved in the bidding would, of course, have to meet specified standards of financial and technical responsibility and of product quality. If the procedure outlined is to serve its purpose, such contract awards must be based strictly on economic criteria rather than on social or political considerations.

2. COMPLETION UNDERTAKINGS

Even with guaranteed price contracts, there may be need for certain supplemental protections for lenders. One need is for the protection of the lenders—and perhaps to some extent, the equity holder—against the risk that the project might not be completed because of such factors as construction cost overruns caused by regulatory delays and other governmental interventions. To furnish some measure of protection against this contingency, **we believe that the government should—in carefully defined circumstances—be able to provide *limited completion guarantees* (for which a fee to the company would be charged) covering a specific number of new synthetic fuel plants with a specified maximum capacity.** Giving firms the opportunity to apply for such a "backstop" financing guarantee is likely to enlarge the number of potential participants in the development of new synthetic facilities.

Such a "completion undertaking" should be designed as a standby arrangement that would in most instances never have to be used. In fact, large construction firms may be willing to operate on fixed price contracts for all or large parts of the plants. Where used, such undertakings would only cover cost overruns above a specified "deductible" level, with fees set in inverse proportion to the amount deductible.

Normally this type of guarantee would only be in effect until the facility becomes operational. The government would stand ready to provide or guarantee any added funds that might become necessary to complete the project in a form capable of supplying the product to be purchased. While

---

[6] The guaranteed price option can only be useful in a regulated setting if the regulatory commissions provide appropriate assurances that allowable rates of return would not be decreased as a result of the guaranteed price contract.

the basic standby fee might be relatively low, the fee for actual use of the guarantee should rise substantially as cost overruns exceeded stated levels; hence, incentives for minimizing cost overruns would be retained.[7] At the same time, the scaled-fee arrangement would discourage excessive use of the guarantee and cause firms to terminate the arrangement once the costs of the guarantee started to outweigh the probable benefits from protection against risks. Firms financially strong enough to provide acceptable completion guarantees to the lenders would, of course, avoid payment of this fee.

3. DEBT GUARANTEES

There may be some circumstances in which relatively efficient firms (prime contractors or subcontractors) cannot take sufficient advantage of the above options and lack a strong enough equity base or special collateral (such as coal reserves) to raise all the funds needed for their part in a project. In such circumstances, an outright government debt guarantee for at least certain parts of a project may be needed to induce a sufficient volume of low-cost bids for the desired synthetic fuel production. This could be accomplished by executing a separate debt guarantee agreement, or by adding to the guaranteed price contract described earlier an assignable provision assuring government payment of at least a minimum total equal to the debt incurred.

* * *

In all of these cases, the equity holders would face the risk of losing their investment. The threat of such a loss should give equity holders a strong incentive to make each project a success.

To permit adequate financing of a sufficient number of the needed first-of-a-kind synthetic fuel projects, we urge that necessary legislative and administrative action be taken promptly that will allow use, where appropriate, of each of the following additional financing options: guaranteed price and purchase contracts; completion undertakings; and debt guarantees. The total amounts of government support to individual participating firms should be held to levels that avoid giving an undue competitive advantage to one type or size of firm.

In conjunction with these financial options, the government should retain the option to buy out and close down the facility, under defined conditions, as an alternative to paying for its completion.

---

[7] To preserve such incentives, it would be important to assure that the government does not directly or indirectly reimburse the firm for the cost of the fee.

CHAPTER 4
IMPROVING ENVIRONMENTAL AND RELATED PROCEDURES

Uncertainty, delay, and sometimes unreasonable restrictions surrounding the application of environmental laws and regulations are other major obstacles to the early approval and construction of new technology plants.[1] Since the enactment of the Clean Air Amendments in 1970 and the Federal Water Pollution Control Act Amendments in 1972, a massive system of legislative and regulatory controls over the environment has evolved. These environmental programs have sound general objectives, such as reduced pollution and the achievement of a cleaner environment, which we fully support. However, the many complex, stringent, and highly uncertain regulatory requirements of these laws pose difficult or impossible barriers to the construction of commercial-size plants utilizing various new energy technologies.

While other energy technologies also have environmental problems, the proposals in this statement are limited to measures to achieve a balanced tradeoff that will allow a limited number of first-of-a-kind synthetic fuel plants to be constructed without undue delay.

In part, the difficulties in applying current environmental regulatory requirements to synthetic fuel production stem from the complexities and uncertainties surrounding the new synthetic fuel technologies. They also stem from the physical characteristics of the likely locations of such plants and the inability of general environmental rules to take into account the nation's

---

[1] Additional obstacles can be posed by various other types of regulations, such as those relating to limitations on oil shale acreage. This statement focuses on environmental restraints, however, in view of the crucial nature of their deterrence to many types of synthetic fuel developments.

need for plants in these locations. Most importantly, they stem from the delays and uncertainties surrounding governmental actions on environmental matters. These can add enough to the already high risks in synthetic fuel ventures to effectively discourage private investment in such ventures.

In our judgment, the difficulties in environmental decision making reflect basic deficiencies in the national mechanism for balancing environmental and energy considerations. Better general methods are needed for

RELEVANT ENVIRONMENTAL RULES IN BRIEF

Present environmental regulations require compliance with a complex layer of federal and state rules governing air quality standards. To preserve and improve the quality of air, National Ambient Air Quality Standards (NAAQS)[1] have been established. The country has been divided into 247 air quality control regions to which these standards are applied. Areas that are cleaner than standards also are subject to rigorous controls "to prevent significant deterioration" of air quality.

The requirements imposed on any new plant are, in part, determined by the classification of the region in which it is to be located. If the plant is to be located in an area which already exceeds air quality standards, the law currently requires that the plant cannot be built unless other polluting facilities in that region are shut down or retrofitted to more than offset the emissions that will be added by the new plant. In addition, the new plant must itself be designed for the "lowest achievable emission rate."

If the proposed plant is to be located in a region whose present air quality is better than the air quality standards, it must (1) use the "best available control technology" and (2) ensure that air quality will not deteriorate in excess of specified "increments" (numerical limitations in maximum allowable increases for specific pollutants over the existing air quality). The requirement for the best available control technology must represent emission limitations based on the maximum achievable degree of reduction for a specific facility, taking into account energy, environmental, and economic impacts in the region. When it is not possible to measure the quantitative levels of emissions accurately, then the Environmental Protection Agency (EPA) may specify the control equipment that will be used. A similar "best available technology" requirement is also provided in the Clean Water Act (Federal Water Pollution Control Act).

1/ The primary AAQS is defined as the level of air quality which the EPA judges necessary to protect the *public health*, allowing for an adequate margin of safety. The secondary AAQS is defined as the level of air quality which is regarded necessary to protect the *public welfare* from any adverse effects.

expediting environmental reviews and resolving environmental disputes. However, in this statement we are not advocating across-the-board changes in environmental procedures. Rather, we are proposing a relatively simple and targeted mechanism for preventing unreasonable delays resulting from environmental disputes over the construction of a limited number of synthetic fuel plants.[2]

The recommended procedures would apply only to specifically designated sites and would modify only those aspects of current environmental laws and regulations that create serious obstacles to the construction of synthetic fuel plants on such sites. These proposals should be designed in a way that does not disrupt the current balance in the relative powers of the Executive Branch, Congress, and the states.

In brief, we recommend consideration of special federal legislation empowering the President to: (1) designate a limited number of specified potential sites for synthetic fuel plants eligible for the provisions of the legislation: (2) expedite governmental decisions concerning possible synthetic fuel plant operations in those sites; and (3) grant limited variances from certain environmental standards and related pollution control technology requirements for those specifically designated sites where national purposes are found to be overriding and where all reasonable alternatives to full compliance with existing environmental laws and regulations have been found to be unworkable.

We believe that these special powers should be carefully circumscribed. **In order to ensure that regional and state interests are safeguarded, this legislation should provide that presidential decisions with respect to any specific site can be overridden by legislative act of Congress or by the state concerned, within a limited time period. The legislation should also provide for prompt, but limited, judicial review of the decisions made. We also believe it is important that the federal legislation for synthetic fuel development be accompanied by companion legislation in the various states where plant sites are expected to be located.**

Our proposal does not challenge the entire environmental controls system but rather seeks expeditious decision making and relief, where needed, for a small number of pioneer plants for new synthetic fuels. We believe that it would, in fact, seldom be necessary to invoke the proposed presidential authority. Subsequent facilities would be expected to meet all applicable environmental requirements.

---

[2] These proposals should also apply to regulations on associated mining operations. In addition, they should be considered for possible future application to other major new energy technologies when those technologies reach the commercial-scale demonstration stage.

More details concerning each of the three main parts of the proposed special legislation are presented in the following sections.

PRESIDENTIAL POWER TO DESIGNATE POTENTIAL SITES

In designating the potential sites for synthetic fuel plants, the President should consider such factors as the availability of water and coal, peat or oil shale, and the costs and convenience of transporting the fuel to end users. As much as possible, these sites should also be capable of supporting synthetic plants with minimum environmental impact. The particular way in which the plants would be financed (whether or not any government support or incentives are involved) should not be a consideration in the designation of sites.

Enough potential sites should be identified to accommodate all interested and qualified applicants willing to build first-of-a-kind synthetic fuel plants and should provide ample opportunity for submission of competing applications. This should establish a "stockpile" of fifteen to twenty-five potential sites from which the President would designate the smaller number of sites that would actually be declared eligible for the special procedures. Sites already identified by potential plant operators should be given priority for inclusion in the stockpile.

To establish such a stockpile, we urge that the Department of Energy promptly begin discussions with energy companies and other interested parties and enlist their cooperation in identifying specific sites for synthetic fuel projects. As soon as the Department of Energy identifies a sufficient number of the potential sites, it should actively solicit competing applications.

EXPEDITING DECISIONS ON SYNTHETIC FUEL PROJECTS

SPEEDY PREPARATION OF ENVIRONMENTAL IMPACT STATEMENTS. Environmental impact statements would have to be prepared before construction of synthetic fuel plants could begin. Present requirements for these statements can involve huge amounts of paper work and a long period of review and comment. These statements also have been subject to extremely protracted administrative and court challenges alleging both incompleteness and inaccuracy. While we welcome recent efforts by the Council on Environmental Quality to cut paper work, reduce delays, and expedite decision making in the preparation and review of environmental impact statements, we believe that special procedures are needed to assure timely environmental decisions regarding synthetic fuel plants.

We urge that for pioneer synthetic fuel plants in specified locations, the President be allowed to place time limits for government action on environmental impact statements and provide effective time limitations on the initiation of litigation over environmental impact statements. Additionally, the special legislation should call for a new form of national environmental impact statement designed specifically for a small number of pioneer synthetic fuel plants, taking into account the inevitable uncertainties connected with such first-of-a-kind installations. A number of specified alternate processes should be included in these impact statements so that backup processes can be substituted without requiring a new environmental impact statement in the event a primary process or equipment proves unsatisfactory.

STREAMLINING PERMIT REQUIREMENTS. Under current laws and regulations, permit requirements for an energy-production plant are numerous and extremely burdensome, ranging from fifty to over one hundred permits from both federal and state authorities.[3] The problem should be mitigated in part by newly instituted Environmental Protection Agency (EPA) procedures for coordinating the necessary reviews with other agencies and expediting decision making. However, the burden of these multi-layer permit requirements can be reduced considerably for first-of-a-kind synthetic fuel plants if the complete cycle (period) of the permit-granting process has a time limit or deadline. The recent EPA regulations governing air quality standards provide that the permitting agency must decide within one year from the date of application whether to grant or deny a construction permit. We welcome strict enforcement of such time limitations and urge their adoption in all permit programs.

For the limited number of pioneer synthetic plants, however, **we recommend that the President be empowered to assure a speedy review process by imposing time limits for the full range of governmental permits required.** Congress, for example, should authorize the President to set a specific deadline for the required government responses anywhere between three months and one year after the date of filing of the necessary papers by the applicant. However, such presidential action to expedite decisions on the permits for specific sites and to achieve the speedier processing of environmental impact statements, described in the preceding paragraph, should

---

[3]/For instance, the typical permits required for a mining operator under the current regulations would include: an environmental assessment report, an environmental baseline study, an environmental impact statement, a state utility site permit, a state mining and reclamation permit, a federal water pollution control permit, a state water use permit, a state water discharge permit, a state air pollution control permit, and approval of a mining and reclamation plan.

be subject to override by Congress and by the affected state government or governments within a specified time and under specified procedures.

As a practical matter, we would expect that the President would usually use a single statement to announce his designation of a particular site, or sites, and his actions to expedite the regulatory processes for site acquisition and for construction and operations on those sites. This would equip both Congress and the state governments with override authority to consider all related aspects of the decisions and to act upon them as a package.

LIMITATION ON LEGAL AND ADMINISTRATIVE CHALLENGES. Resolution on environmental disputes has proved time-consuming and costly. Ambiguous and repetitive litigation and administrative reviews have often been the major causes of delays in the construction of energy production facilities. To ensure expeditious decisions on environmental disputes, time limitations on court challenges and administrative reviews might be specified along the lines provided in the Alaska Pipeline legislation. (The Alaska Pipeline Act provided a sixty-day limitation from the date of enactment for court challenges to be brought on environmental grounds.)

We recommend that the proposed special legislation impose a time limit on the initiation of any litigation to be brought in opposition to the permit approval for any of the designated synthetic fuel plants, and that no legal or administrative challenges be permitted beyond that date unless it can be shown that the actions involved are unconstitutional or that the facility fails to comply with the provisions of the proposed special legislation. These limitations would eliminate most of the time-consuming and costly delays caused by protracted litigation and administrative review. This would force the parties involved to consider all the relevant factors nearly simultaneously, thereby facilitating a balanced initial decision and foreclosing further disputes.

PRESIDENTIAL ENVIRONMENTAL VARIANCES

In the event that the special procedures we have recommended for speedy decision making are not adequate to assure timely construction of all the synthetic fuel plants that are needed in the national interest, an additional mechanism should come into operation. **We recommend that in such cases, the President be empowered to grant limited variances from environmental standards and related requirements for specific projects in specified areas.** This authority could be exercised through Executive Orders, but should be subject to reversal by both Houses of Congress within a specified time limit (possibly ninety days). His decision should also be subject to speedy judicial review limited to constitutional issues and to determining whether the Presidential Order covers the types of variances authorized in the legislation. Finally, it would be advisable to allow states to veto, within stated time limits, the presidential "variance" decision as far as their own territory is concerned.

While use of a special presidential variance normally would be invoked only after the procedures under the speedy decision-making process have been exhausted, the President should in exceptional cases be empowered—and should act—to grant limited environmental variances from the outset to specified synthetic fuel projects in a particular area, if this can be shown to be in the national interest, and if it is clear that the plant cannot be built without such variances.

The President should also be empowered to grant this variance after an approved synthetic fuel plant goes into operation, if its emissions turn out to exceed the expected or approved levels despite best efforts to the contrary.

In suggesting that the President have a limited authority to grant variances from prevailing standards, we are particularly concerned with those cases where plant designers believe that their plant will meet environmental requirements but where they cannot offer sufficient assurances on that point to satisfy would-be lenders.

To preserve some incentive to reduce emissions in these cases, the President could be given the power to levy reasonable emissions fees or effluent charges on the excess of the plant's emissions over the regular standards. By "reasonable" we mean a charge that is not trivial but also is not punitive and not disproportionate to the rate of return being earned by the plant. CED has long felt that the use of such economic incentives has many advantages over rigid regulatory standards and should be tried to the extent practicable.[4] We believe that the use of appropriate economic incentives in this kind of case, as an alternative to detailed regulation, could achieve an appropriate balance between considerations of equity and the environment.

VARIANCES FROM AIR QUALITY REQUIREMENTS. A critical difficulty in meeting air quality requirements is that they make no clear distinction between natural and man-made pollutants. For example, in some lightly populated areas the level of natural dust or the amount of natural hydrocarbons given off by trees may combine with pollutants transported from far away to cause the overall level of ambient pollution to exceed present air quality standards. Pollutant concentrations of particulates and ozone in

---

[4] See *More Effective Programs for a Cleaner Environment*, CED Statement on National Policy (1974).

parts of the undeveloped oil shale regions of Colorado and Utah are re-
ported to frequently exceed the ambient air quality standards. Since there
are no local controllable emissions that could be discontinued to offset new
emissions from oil shale operations, strict enforcement of the ambient air
quality standards for particulates could create an absolute barrier to oil shale
development in those areas.

Compliance with the air quality standards has been clarified somewhat
with the issuance of a June 1978 EPA regulation on prevention of significant
air quality deterioration. The new regulation excludes "fugitive dust" in as-
sessing the air quality impact of a new plant and in calculating allowable
emission increment limitations. At present, this exclusion applies only to
one kind of pollutant, particulate matter, and there is no guarantee that the
exclusion provisions will be continued or extended to other kinds of pollutants. We believe, therefore, that with respect to this proposed small number of synthetic fuel plants, the President should also be empowered to grant
variances, subject to congressional override, on compliance with air quality
requirements for particulates and for other kinds of pollutants (such as
ozone and nitrogen oxides).

GREATER CERTAINTY IN POLLUTION CONTROL REQUIREMENTS.
One of the most debilitating uncertainties for a prospective builder of a synthetic fuel plant is the possibility that applicable "best available control
technology" requirements would be tightened after his plant was already
built. This could raise costs immeasurably, shut down the plant, or both. We
believe that appropriate action to minimize this uncertainty can do much to
encourage private construction of pioneer synthetic fuel plants, without disproportionate cost or damage to the public interest.

**Therefore, in cases where a pioneer synthetic fuel plant has installed
an acceptable emission control technology but the authorities subsequently decide that retrofitting such a plant is necessary to remove pollutants found to be more harmful than originally thought, or to meet more
stringent control standards, we recommend that the costs of such retrofitting should—in the case of this limited number of designated plants—be
borne by the government requiring such retrofitting.**

This issue will be clarified when EPA issues its New Source Performance Standards for synthetic fuel plants. These emission standards represent
limitations on the maximum allowable emissions of specific pollutants
(such as sulfur dioxide and particulates) applicable to specific categories of
plants and processes. Although EPA has not yet developed specific numerical levels of new source performance standards for synthetic fuel plants, it
recently did issue guidelines for controlling emissions for Lurgi coal gasifi-

cation plants.[5] These guidelines are intended to assist state, local, and regional EPA offices in determining "best available control technology" requirements for Lurgi coal gasification plants on a case-by-case basis. We hope these guidelines will serve as the basis for timely development of technical definitions of new source performance standards as well as "best available control technology" requirements for coal gasification plants in the future.

The construction of a limited number of pioneer plants as proposed in this statement would provide the necessary experience to set intelligent emission standards and choose optimal control technology. **As soon as experience with these pioneer plants permits, we urge that EPA establish realistic new source performance standards for synthetic fuel plants, incorporating the available control technology that best balances the public interest in environmental control, energy production, and economic performance.**

WATER SUPPLY AND WATER QUALITY

Production of synthetic fuels from coal and oil shale could require large amounts of fresh water and could result in discharge of large quantities of polluted water. This would pose a serious problem because much of the low-sulfur coal mines and almost all of the high-grade oil shale are located in arid or semi-arid western regions.

However, it may be that these problems can be minimized by the use of closed systems in which the water would be continually recirculated within the plant and would require only process water and limited "make-up" additions, thus avoiding discharge of waste water and minimizing the net water requirements.

Construction of these first-of-a-kind synthetic fuel plants would certainly provide more accurate information regarding future water requirements for larger-scale production. We believe that judicious selection of plant sites, better management of overall water resources, and the application of appropriate production technology that will minimize water consumption and water pollution would resolve many of the concerns associated with the problems of water supply and water quality.

---

[5] U.S. Environmental Protection Agency, *Control of Emissions from Lurgi Coal Gasification Plants* (March 1978). The new source performance standards for coal-fired power plants that were established by EPA in May 1979 also provide some exemptions from the standards for the first few plants to be constructed that use any of the following four emerging technologies: solid solvent refined coal, atmospheric fluidized bed combustion, pressurized fluid bed combustion, and coal-derived liquids.

CONCLUDING REMARKS

We believe that the proposal for special environmental procedures with respect to the designated synthetic fuel plants should be widely discussed by the public and interested parties for a reasonable period prior to legislative action. Numerous studies by federal agencies on the possible environmental impact of synthetic fuel production have already been completed or are currently in preparation.[6] **To encourage public involvement, we urge that the appropriate congressional committees begin hearings on this proposal promptly and invite extensive representation of all interested points of view.** In these hearings, we urge both government and industry to be as explicit as possible about the potential locations and operating characteristics of the proposed synthetic fuel plants. These hearings should provide ample opportunity to clarify and resolve a large share of the existing uncertainties and concerns about the environmental impact of synthetic fuel production.

---

[6] For example, the Energy Research and Development Administration (ERDA), now of the Department of Energy, completed its initial assessment of the environmental impact of synthetic fuel development programs (*Alternative Fuels Demonstration Program: Final Environmental Impact Statement*) in September 1977.

MEMORANDA OF COMMENT, RESERVATION, OR DISSENT

Page 2, by CHARLES B. BOWEN, JR., with which ROBERT B. SEMPLE has asked to be associated

Either our energy situation and its effect upon our vulnerability to foreign political and economic pressures is becoming a near-term emergency or it isn't. If it isn't, then why should we not let things take their normal course and let the synfuel technologies compete in terms of economics and environmental impact?

If, as I believe, it is a growing emergency, this program is far too limited and leisurely, with its tacit acceptance of modest reductions in time-consuming and unproductive congressional, state, and regulatory reviews after the program has been adopted.

This emergency has been created primarily by the paralysis of national and local political leadership which has devoted more to pursuit of negatives than positives. It has been focused more on why needed things cannot and should not be done rather than how to do them quickly and properly.

To break out of this paralytic mold, we need a Federal Synfuels Corporation whose sole objective should be to secure a significant privately-owned and operated initial synfuel production capacity at a reasonable cost and with a practical minimum of adverse environmental impact by construction of at least two commercial-scale plants of each of the competing processes.

The Corporation should have whatever authority is necessary for the prompt execution of its task, including encouragement of private financing of the plants by any or all.of the appropriate methods outlined in the paper. Its authority and limitations should be settled between the executive and the legislative branches *before* the program starts and be legislatively protected from interminable judicial and administrative delays. It should be a project-oriented organization, not contaminated by organizational association with the present Department of Energy. Except for last-resort purchasing or subsidy functions, it should by law terminate its activities within two years of on-stream operation of the last plant.

It is not clear why the potential of a few billion dollars extra in one-time capital costs or overruns should concern us much when our foreign suppliers have demonstrated that in a few months via arbitrary price increases they can increase our energy costs that much and more on a continuing basis.

It seems more useful to get going with a synfuel program clearly in the national interest and to pay the billions to ourselves rather than to worsen our balance of payments still further by paying it to them.

Page 2, by GEORGE C. McGHEE, with which ROBERT R. NATHAN has asked to be associated

I heartily approve the concept of the CED paper *Helping Insure Our Energy Future: A Program for Developing Synthetic Fuel Plants Now.* I believe it will make an important contribution toward solving our nation's energy problem. I would, however, have preferred a higher goal for the aggregate output of the various synthetic projects recommended, at least to the two million barrels a day level for synthetic hydrocarbons by 1990 as proposed by President Carter in his speech of July 15. Our shortage of liquid hydrocarbons by that time will be even greater.

Page 3, by JACK F. BENNETT

In my judgment, the estimate of $15 billion for ten synthetic fuel plants is too low.

Page 3, by G. BARRON MALLORY

There must be an immediate deregulation of gasoline prices and a reduction of the price controls on fuel oil, diesel, gas and other sources of energy in order to permit the market to establish prices. The policy statement should take a position in opposition to the so-called "windfall profits tax." Otherwise, the oil companies and financial institutions will probably not have enough capital to invest in facilities and the development of synthetic fuels. The "windfall profits tax" would leave little, if any, capital in order for the private companies to invest the balance of the $15 billion over the next decade.

Pages 3 and 17, by G. BARRON MALLORY

The policy statement is excellent in many respects, but there is a serious deficiency. References are made to the "all-events tariff." Probably this term or phrase has a meaning only for insiders and most of the readers of the statement will be confused. "Tariff" in Webster's dictionary is said to mean a system of duties imposed by a government on imported goods or a duty imposed on imports. "Surcharge" would be a better term.

Page 4, by ROBERT R. NATHAN, with which ROBERT B. SEMPLE, HOWARD S. TURNER, and FRAZAR B. WILDE have asked to be associated

Substantively this is an excellent policy statement which should have a constructive impact in a highly important area at a critical time. Regrettably it was not available some months ago.

The positive major thrust of the statement offsets some degree of timidity concerning the role of government. The objectives realistically relate to the nation's critical energy needs. However, the words often seem designed to placate those who look with distress upon any governmental role or responsibility.

For instance, it is suggested that the government would pay perhaps no more than around $2 billion out of the estimated $15 billion total cost of these synthetic plants. Both numbers will very likely prove to be unrealistically low, especially the government's portion. The government participation should be no larger than necessary, but emphasis should focus on what is needed to make the program a success and not on some maximum level of government funding.

Most distressing are the comments concerning the principles for public-private cooperation. It is stated that the government should assume only risks and costs that cannot be assumed on strictly commercial grounds, such as those caused by government actions. This fails to emphasize that there are serious burdensome technological risks. In fact, technological and cost uncertainties are the principal reasons why an intelligent and constructive governmental program is needed. It is self-defeating to start with the assumption that all risks except those deriving from governmental restraints and actions can and will be assumed on strictly commercial grounds.

If this principle were to be strictly construed the prospects of decreasing our energy dependence and increasing our energy independence would be dismal indeed. Government ownership and operation or detailed governmental interventions should, of course, be minimized. But we must recognize that an important governmental role is inevitable in such a major effort. This statement generally proposes sound policies and programs consistent with this urgent national need and it is to be hoped that some of the wording will not be used by the fainthearted to fight against these sound policies and programs.

Page 7, by C. WREDE PETERSMEYER

I think this is a practical, business-like approach to developing synthetic fuels. I am hopeful that the program recommended by President Carter in his address to the nation on July 15, which involved massive expenditures of government funds in a large-scale crash program for producing synthetic fuels under a governmental authority, will be modified along the lines of this statement, i.e., building a small number of first-of-a-kind operating plants to demonstrate the commercial feasibility of converting coal and shale to synthetic fuels, to rely primarily on the private sector, and to minimize the net expenditure of government funds.

Page 10, by W. D. DANCE, with which H. J. HAYNES and JAMES Q. RIORDAN have asked to be associated

I would add to this list "enhanced secondary and tertiary recovery methods from existing oil fields." This option would provide the quickest realizable gains in production. The government could encourage such recovery by price decontrol and exemption from the "windfall profits" tax.

Page 17, by ROGER B. SMITH

I believe the "all-events tariff" is inconsistent with the general objectives of the economic incentives stated throughout this report. Also, caution is urged regarding the recommendations for loan guarantees and price contracts. Such measures should be approached with caution because of the potential for further disruptions of market incentives for energy development.

Page 18, by J. W. McSWINEY

The general concept of the "all-events tariff" has recently been considered and rejected in a ruling by an administrative law judge for the Federal

Energy Regulatory Commission (June 6, 1979, CP78-391, Great Plains Gas-ification Associates, R. N. Zimmet, Administrative Law Judge). The judge indicated that a high projected rate ($5.56-8.21 per MCF of synthetic gas) should not be borne by the customers of a single gas distributor or even of a consortium. He also indicated that these customers should not bear the capital costs in the event that the plant were a failure. Instead, such extraordinary costs should be shared by the country as a whole, since all would benefit.

I submit that this ruling would indeed appear to be a "correct" position, logically.

Page 20, by JACK F. BENNETT

The fact that a project might produce synthetic fuel which would be transported in a regulated pipeline does not mean that the project itself must be subject to regulation. Projects for the production of conventional oil have not normally been subject to utility regulation, even though transportation of the oil was to be through a regulated pipeline. Moreover, the Natural Gas Policy Act of 1978 provides for deregulation of the prices of new natural gas in 1985.

APPENDIX

GAS AND OIL FROM COAL AND OIL SHALE

There are several technologies for producing gas from coal and oil from shale now ready for commercial use. There are other technologies for producing liquids from coal and gas from oil shale and coal that are under intensive pilot plant development and are likely to be ready for commercialization within one to three years.

THE STATE OF TECHNOLOGY

Medium BTU gas (450-650 BTUs per cubic foot) was produced for 100 years in urban plants for local distribution. Technologies developed in Germany during World War II resulted in improvements in coal gasification. More recent tests indicate that the state of technology is advancing rapidly enough to permit gasification and subsequent processing of coal to produce "pipeline" quality (1000 BTU) gas that is equivalent to natural gas. Such gas can be directly introduced into existing nationwide pipelines transporting natural gas and commingled with that gas.

It is also likely that less costly medium BTU gas can be produced from coal for efficient use near the generating plant, and that numerous industrial gas users, now dependent on 1000 BTU gas, could adjust to using medium BTU gas. Gas for fuel use can also be produced from oil-containing shale, although most interest in this raw material has centered on the production of a substance that is subsequently refined for use as a liquid. Shale containing ten to thirty-five gallons of crude oil per ton exists in enormous quantities, principally in Colorado, Utah, and Wyoming.

Production of petroleum-type liquids from coal was accomplished in Germany during World War II. South Africa relies on the World War II methods, improved through twenty years of operating experience and engineering, in the large SASOL installations now being expanded in that country. Some methanol was produced from coal in the United States until the early 1950s, and improved processes for making it from coal have been developed since then. Also, a number of more direct processes for coal liquefaction are now well advanced on a pilot plant scale.

THE NEED FOR FIRST-OF-A-KIND COMMERCIAL PLANTS

Energy companies are prepared to build first-of-a-kind commercial plants for producing gas and oil from coal and oil shale if regulatory and environmental obstacles and economic uncertainties can be overcome. Since these technologies are all based on the use of plentiful domestic energy resources, coal and oil shale, they offer an early opportunity to expand domestic energy production and to provide a resource base on which to build a major domestic energy industry as the need develops.

Oil shale resources, which are not being used now, can only be tapped by processes for separating the substances which can be converted to oil or gas from the sand and rock in which they are locked.

Although coal can be burned directly to produce heat and electrical energy, gasification and liquefaction offer significant advantages:

- The existing interstate network of pipelines can be used to transport the oil and high BTU gas manufactured from coal or oil shale. Pipeline transportation is, in general, the cheapest mode of moving hydrocarbon energy.
- At the point of end use, air pollution from burning this substitute gas and oil will probably prove to be substantially less than pollution from present methods of burning coal.
- These processes can provide critically needed liquid fuels for cars, trucks, and other forms of transport.

Each of the proposed synthetic fuel plants would use a technology that is ready for scale-up to commercial-size levels. Each of these plants should be a single production unit large enough to bring unit product cost down to the point on the cost curve beyond which only relatively limited economies of scale could be realized, and from which engineering and cost estimates of future larger plants can be made with high confidence. Two plants based on different technologies for each major type of synthetic fuel, and each operated by a separate company, is probably the minimum number needed to provide realistic and reliable results. Reliance on a single plant using a single technology, a single coal or oil shale source, and a single technical management is too large a risk to accept in light of the great importance and urgency of proving large-scale commercial feasibility.

Opinions vary on what the size of each of these plants should be in order to accomplish the program objectives. Smaller units will have higher costs per unit of output, and therefore, will be less economical and less able to demonstrate full-scale production costs and environmental performance. Plants so small that the unit costs of fuel produced would be two to four times the price of comparable conventional fuels would not provide convincing evidence of the expected costs of plants for optimum production;

because of their high costs, they would either have to be abandoned after a demonstration period or expanded, if possible, to full optimum size.

Accordingly, it appears that small plants would be unlikely to elicit the necessary private financial resources and the commitment of the private energy industries. On the other hand, very large plants can involve capital costs which are so high that they may be a major impediment to actual construction.

HIGH BTU GAS FROM COAL (PIPELINE QUALITY)

Projects for high BTU gas production from coal, which five private groups currently have in the design stage, range in capacity from 125 to 288 million cubic feet per day (22 to 50 thousand barrels per day oil equivalent). The Gas Research Institute has estimated that the unit cost of gas from plants producing 50 million cubic feet per day will be about 30 percent greater than the cost of gas from a 125 million cubic feet per day plant, while doubling the size of the plant to 250 million cubic feet per day might further reduce unit costs by 20 to 25 percent. Capital cost estimates for the larger plants now being proposed range from $900 million to $1.3 billion each (excluding the associated coal mines). It should be noted that eastern and western coals differ and, therefore, require quite different process treatment.

MEDIUM BTU GAS FROM COAL

Medium BTU gas is suitable for industrial heating, for power generation, and for feedstock for petro-chemical processes. It cannot be economically piped as far as high BTU gas and requires a separate pipe system because it cannot be mixed with high BTU gas. Therefore, its use will probably be restricted to relatively small geographic regions containing both suitable coal and a concentration of industrial users sufficient to consume the output of an economically-sized plant.

The cost per BTU should be less than that of high BTU gas because the step of upgrading energy content, necessary in high BTU gas production, is not required. However, the added costs of building separate pipe networks and of storage may, in some cases, offset the lower production costs. Thus, the market for medium BTU gas may be limited, but it is important to gain experience now by constructing a limited number of medium BTU plants where favorable geographical and market conditions exist.

SHALE OIL

Shale oil provides a huge potential for domestic oil and gas. Three separate shale oil plants probably will be needed in order to prove the production technologies, economics, and environmental control technologies. It is particularly important to produce liquids from shale because of this nation's very large requirements for petroleum products for cars, trucks, and other mobile power sources, and because the United States now meets almost half of these requirements through imports.

Oil shale is a sedimentary rock containing varying amounts of solid organic material called kerogen. When heated to 900°F., the kerogen decomposes into hydrocarbons and carbonaceous residue. The hydrocarbons, when cooled, condense into a liquid called shale oil. The crude shale oil is then refined to produce various petroleum products.

Production of shale oil involves mining and breaking up the oil shale for a surface retorting process or cracking and retorting *in situ* (underground) process. Surface retorting is generally thought to be ready for commercialization, but retorting *in situ* is still in the development stage. Shale oil has been produced abroad on a small scale for a number of years. There has been intermittent production of shale oil in the United States for more than one hundred years, but it has thus far not been sufficiently competitive with natural petroleum to result in commercial production. Nevertheless, of the various new technologies for producing synthetic liquid fuels, shale oil is among those closest to commercialization.

U. S. resources of high quality shale oil (over 25 gallons per ton) have been estimated at 600 billion barrels—more potential energy than is contained in the combined proven U. S. reserves of oil and natural gas. The principal barriers to the commercialization of shale oil are release of mining rights, uncertainties regarding price, environmental constraints, and, to date, the inability to compete economically with imported oil. Present estimates of the capital cost of a minimum production-sized plant are in the order of $1 billion per 50,000 barrels a day capacity.

Environmental considerations are a major constraint on shale oil production. The richest shale oil resources are concentrated in the semi-arid, pristine areas of Colorado, Utah, and Wyoming. Air quality constraints in these regions, some of which are classified as non-attainment areas and may preclude the development of this resource, tend to limit the size of the industry in any one area. Water availability for shale oil production is another potential constraint which can limit the extent of the ultimate development.

COAL LIQUEFACTION

Coal liquefaction could significantly reduce this nation's liquid fuel problems, although this technology is somewhat behind coal gasification in terms of commercial readiness. The product of coal liquefaction can be used directly as a boiler fuel, or it can be upgraded by refinery type processes to produce clean products.

Although the technical feasibility for producing liquid fuels from coal was demonstrated in Germany during World War II, the only commercial facility in the world producing liquids from coal is the SASOL plant in South Africa, which has been operating for over twenty-five years. In this facility, coal is gasified to a low BTU gas of specified composition. This gas is then converted into a liquid fuel through a technology known as the Fischer-Tropsch process. Such gas from coal can also be converted to methanol, which is itself a versatile fuel that can be further converted into high octane gasoline.

Several large pilot plants for direct conversion of coal to liquids (using different technologies) will soon be completed in the United States. One of these plants, based on a technology developed by Hydrocarbon Research, Inc. and known as the H-Coal process, is scheduled to be completed in 1979 under the joint sponsorship of the Department of Energy and a private industry consortium. The second one, the Exxon Donor Solvent pilot plant, is expected to be completed by early 1980. The third, a liquid solvent refined coal demonstration plant (SRC II) developed by the Gulf Oil Company, is planned to be completed in 1983. These and other processes are being tested in pilot plants and are expected to provide much-needed information about commercial scale-up possibilities.

SOLIDS FROM SOLVENT REFINED COAL

While this statement has focused on synthetic liquid and gaseous fuel, a related coal conversion technology is the coal solifaction process, converting coal into synthetic solid fuel. The solifaction process is somewhat similar to the first step in coal liquefaction techniques (dissolving coal with a solvent). The solvent refined coal (SRC) process is the only solifaction process which is being actively developed at present. This process involves dissolving pulverized raw coal in a solvent under high temperatures and pressures. The dissolved product is then filtered to remove impurities. The end product is considerably lower in ash and sulfur content than the original coal and has a high heating value. The product remains as a solid fuel at room temperatures, but it can easily be liquefied. The SRC process is, therefore, being developed as part of coal liquefaction efforts. The SRC technology has large number of potential near-term applications, but it has not received as much attention as other processes. At present, there are two pilot plants operating in the United States. The major difficulties with the solifaction process are its high filtration costs and the development of handling methods for the solid product.

OBJECTIVES OF THE COMMITTEE FOR ECONOMIC DEVELOPMENT

For thirty-five years, the Committee for Economic Development has been a respected influence on the formation of business and public policy. CED is devoted to these two objectives:

To develop, through objective research and informed discussion, findings and recommendations for private and public policy which will contribute to preserving and strengthening our free society, achieving steady economic growth at high employment and reasonably stable prices, increasing productivity and living standards, providing greater and more equal opportunity for every citizen, and improving the quality of life for all.

To bring about increasing understanding by present and future leaders in business, government, and education and among concerned citizens of the importance of these objectives and the ways in which they can be achieved.

CED's work is supported strictly by private voluntary contributions from business and industry, foundations, and individuals. It is independent, nonprofit, nonpartisan, and nonpolitical.

The two hundred trustees, who generally are presidents or board chairmen of corporations and presidents of universities, are chosen for their individual capacities rather than as representatives of any particular interests. By working with scholars, they unite business judgment and experience with scholarship in analyzing the issues and developing recommendations to resolve the economic problems that constantly arise in a dynamic and democratic society.

Through this business-academic partnership, CED endeavors to develop policy statements and other research materials that commend themselves as guides to public and business policy; for use as texts in college economics and political science courses and in management training courses; for consideration and discussion by newspaper and magazine editors, columnists, and commentators; and for distribution abroad to promote better understanding of the American economic system.

CED believes that by enabling businessmen to demonstrate constructively their concern for the general welfare, it is helping business to earn and maintain the national and community respect essential to the successful functioning of the free enterprise capitalist system.

CED BOARD OF TRUSTEES

HONORARY TRUSTEES

CED COUNTERPART ORGANIZATIONS
IN FOREIGN COUNTRIES

Close relations exist between the Committee for Economic Development and independent, nonpolitical research organizations in other countries. Such counterpart groups are composed of business executives and scholars and have objectives similar to those of CED, which they pursue by similarly objective methods. CED cooperates with these organizations on research and study projects of common interest to the various countries concerned. This program has resulted in a number of joint policy statements involving such international matters as energy, East-West trade, assistance to the developing countries, and the reduction of nontariff barriers to trade.

CE	Círculo de Empresarios *Serrano Jover 5-2, Madrid 8, Spain*
CEDA	Committee for Economic Development of Australia *139 Macquarie Street, Sydney 2001,* *New South Wales, Australia*
CEPES	Europäische Vereinigung für Wirtschaftliche und Soziale Entwicklung *Reuterweg 14,6000 Frankfurt/Main, West Germany*
IDEP	Institut de l'Entreprise *6, rue Clément-Marot, 75008 Paris, France*
経済同友会	Keizai Doyukai (Japan Committee for Economic Development) *Japan Industrial Club Bldg.* *1 Marunouchi, Chiyoda-ku, Tokyo, Japan*
PSI	Policy Studies Institute *1-2 Castle Lane, London SW1E 6DR, England*
SNS	Studieförbundet Näringsliv och Samhälle *Sköldungagatan 2, 11427 Stockholm, Sweden*